Overlooked Causes and the Prevention

The Prevention

Cancer, Cardiovascular Heart Disease, Multiple Sclerosis

Arthur Douglass Alexander III

authorHOUSE®

AuthorHouse™
1663 Liberty Drive
Bloomington, IN 47403
www.authorhouse.com
Phone: 1-800-839-8640

First published by AuthorHouse 2/25/2010

ISBN: 978-1-4490-5356-7 (e)
ISBN: 978-1-4490-5357-4 (sc)

Library of Congress Control Number: 2009912610

Printed in the United States of America
Bloomington, Indiana

This book is printed on acid-free paper.

Contents

DEDICATION

To the many patients who have suffered from life-threatening chronic diseases
and
had the courage to seek life-saving non-toxic alternative &
complementary therapies.

ACKNOWLEDGMENTS

I give special acknowledgment to the following colleagues, friends and family who have played such an importannt role over the past 30 years in establishing and reinforcing my belief in the basic tenet of the Hipocratic oath – "To heal and do no harm" – and in the research and discovery of medical therapies that permit the successful, non-toxic treatment of serious life-threatening diseases such as cancer, multiple sclerosis and cardiovascular disease:

My mentor from 1975 until her death in 1990, the late Dr. Virginia C.W. Livingston, MD, whose brilliant clinical and medical research led to the discovery that a mycobacterium is the primary cause of most cancers.

My mentor and colleague, Dr. Hans A. Nieper, MD, who developed a whole class of non-toxic compounds that have shown great efficacy in the treatment of cancer and multiple sclerosis (MS). We met initially at Sloan-Kettering Institute (New York) in 1959 where Dr. Nieper was a post-doctoral fellow and I was the executive assistant to the Scientific Director. Our friendship and collaboration continued until his death in 1999.

The late Dr. C. Chester Stock, PhD, the Scientific Director of Sloan Kettering Institute for Cancer Research, who trained me in the biochemistry and screening of new potential chemotherapy agents.

My mentor and colleague, Dr. John J. Majnarich, PhD, President and Scientific Director of BioResearch Laboratories in Redmond, Washington; Director of Livingston Foundation for Cancer Research; close friend, colleague and Scientific Advisor to the late Dr. Virginia Livingston, MD; and co-holder with Dr. Livingston of numerous key patents in the field of cancer microbiology and immunology.

My daughter, Nancy J. Kennelly, whose patience in tirelessly proof-reading this manuscript together with many helpful suggestions to clarify my writing for the benefit of patients and lay readers.

INTRODUCTION

This book will attempt to expose, delineate and document the sometimes over- looked and often ignored causes of three major life-threatening chronic diseases – CANCER, CARDIOVASCULAR DISEASE and MULTIPLE SCLEROSIS – as well as providing a therapeutic basis for their prevention. Not unlike the recent monetary system meltdown, the cost of treating these diseases borders on bankrupting medicare, medicaid and the private medical insurance companies.

Cancer

Cancer reaches into every family around the world; it has become an epidemic and spares no one. One person in the United States dies of this dreaded disease every minute, over 1,400 a day, and over 550,000 a year. The approved, conventional treatments – surgery, radiation and chemotherapy – offer no real assurance of cure. The financial burdens these place on the patients, their families and insurance companies are devastating. Many billions of dollars a year are spent in the United States on cancer treatments – again without success or promise of cure. Proportionately, little money is spent by the large pharmaceutical companies on seeking the root causes of cancer, and on effective preventive therapies that can reduce the growing cancer-ridden population in the United States. Let me explain one of the primary reasons this came about.

In 1945 Dr. Cornelius Rhoads, MD, working at the Chemical Warfare Center at Fort Detrick, Maryland, discovered that nitrogen mustards (a class of highly toxic chemicals, including World War 1 mustard gas) would kill cancer cells, and also most other non-cancerous cells. Armed with this knowledge, Dr. Rhoads approached Alfred Sloan, head of General Motors and Charles Kettering, engineering genius who invented and perfected the automobile starter motor asking them to put up some 6 million dollars to construct a multi-disciplinary cancer research center adjacent to the Rockefeller's Memorial Hospital in midtown Manhattan. He convinced them he would find the "magic bullet" that would eliminate the scourge of human cancer just as Fleming in the 1930's had discovered penicillin to cure infection.

On completion of Sloan-Kettering Institute for Cancer Research, Dr. Rhoads contacted the leading pharmaceutical companies, universities and outstanding researchers asking them to submit their potential anti-cancer agents under confidentiality agreements to SKI for testing at no cost against three primary cancers in mice: Sarcoma S180, Adenocarcinoma Ca755, and Leukemia L1210. Small segments of these cancers were gently placed under the back skin of the test mice with a trocar (a hollow syringe) and allowed to vascularize and grow for a day.

Usually 10 cages, containing 10 mice each were used to test each new agent against each of the three primary screens. One or two cages of the mice were controls and received no agent injections, allowing the tumor segments to grow unimpeded, while the remaining mice in the first cage received a standard low dose of the new agent. The mice in each subsequent cage received an increased dose until the mice in the last cage received a maximum dose. The daily tumor growth in each mouse was carefully measured. After a week of testing the effectiveness of the test agent to inhibit the experimental cancers was calculated – if the agent in one or more of these animal cancers inhibited growth of the tumor by 25% or more, the agent was then processed through up to 108 additional animal and cell culture systems to carefully define the breadth of activity and possibly the mechanism of inhibitory action. In the primary screening, the LD50 (the dose level; that killed 50% of the test mice in that cage) was also noted. The company received a confidential report and was allowed to patent the new chemotherapeutic agent and to begin limited pre-clinical trials, and finally broad clinical application in humans.

Cancer Chemotherapy Defined

The therapeutic concept of cancer therapy is based upon the fact that cancer cells have a voracious appetite to sustain their rapid growth; consuming four to five times the amount of nutritional foods (metabolites) consumed by normal cells.

With this knowledge, normal metabolites are altered very slightly to render the metabolite a poison; thus, as the cancer cells compete for the poisoned metabolite, four to five cancer cells are killed for every normal cell. Examples of classical chemo- therapy agents are 5-fluorouracil (5-FU), in which a single fluorine atom is added to the metabolite uracil, and 6-mercaptopurine, in which a –thio (-SH) group is added to the purine metabolite.

Often, after a sustained period of weekly infusions of a chemotherapeutic agent, a patient's cancer cells become resistant to that particular agent, requiring the oncologist to select a different agent for continued therapy. Most of these agents are highly toxic and the natural disease-fighting immune system of the patient becomes extremely weakened with time. This is the reason immune-supportive nutrition and supplements are so critically important during the use of cancer chemotherapy.

Sloan-Kettering screening of potential chemotherapeutic anti-cancer agents is funded by the National Cancer Institute, the American Cancer Society, the Damon Runyon Foundation, Mary Lasker and many others. It is interesting to note that one of the most widely used chemotherapy agents, 5-flurouracil (5-FU), was tested at SKI under American Cancer Society funding. The ACS collects a 50% royalty on each milligram of 5-FU sold world-wide. Cancer chemotherapy equates to megabucks for the pharmaceutical houses. This is one of the primary reasons why these houses, working with the Food and Drug Administration (FDA), resist approving non-toxic, low-cost alternative and complementary therapies such as vaccines that show promise of cure, for they constitute a major threat to the continuation of the present chemotherapy industry.

A growing number of cancer patients are seeking alternative and complementary treatment options that are less traumatic, non-toxic immune enhancing therapies. Unfortunately many of these are not approved and most medical insurance will not pay for them. Thus only the financially able can seek these therapies, and often must go to Mexican or European clinics where these are available.

Cardiovascular Disease

Current clinical practice is focussed on reducing a build-up of lipids and plaques (cholesterol and fatty deposits) from arterial walls which frequently cause these walls to narrow (arteriosclerosis) often resulting in heart attacks and strokes. Recent studies have implicated a relationship between cardiovascular disease and inflamation. The newest, highly sensitive CRP (Commit-Reactive Protein) test can predict risk for cardiovascular disease, heart attack and stroke. Cholesterol is still an important marker, but one-half of those with low cholesterol levels still suffer hearrt attacks and strokes due to inflammation.

A number of clinical researchers in England reported finding strange microbial bodies in the hearts of patients who had died of coronary disease. Drs. Virginia Livingston and Alexander-Jackson noted that microbial bodies were present in the lesions of heart disease and particularly numerous where blood vessels had ruptured.

Drs. Joseph Melnick and Edith Speir of New York proposed a link between the P-53 protein, arterial disease and human cancer. Dr. Melnick further confirmed the role of cytomegalovirus in blocking the action of the vital growth-limiting P-53 protein

Multiple Sclerosis

Multiple sclerosis is a progressive, degenerative disease of the central nervous system. It involves destruction of the myelin sheaths around nerve axons, resulting in scar tissue plaques and ultimately destroying the nerves and their ability to function. MS patients experience membrane deterioration, *sclerosis*.

There appear to be several primary causes of MS including exposure to aluminum, dairy products, heredity and viruses. MS of all chronic, life-threatening diseases is perhaps the the most difficult to study and to follow clinically. It is often changeable and highly variable in its symptoms and progression through the patient's body. Physical activity becomes highly impaired and often declines from only mild impairment initially, walking with minor assistance but can drive; to moderate, walking only short distances with assistance; to moderately severe, requiring a wheelchair, but can still stand; to severe, requiring a wheelchair or totally bedridden.

Orthodox therapies for MS have consistently proven ineffective over any reasonable period of time, and in many cases highly toxic. Alternative orthomolecular therapies, though generally unavailable in the United States, have been shown to be more effective and well tolerated without toxic side effects. These will be discussed along with their mechanisms of action.

1. The Role of Certain Mycobacteria in Causing Cancer

Background

Cancer reaches into every family around the world; it spares no one. In the United States today more than 1,400 persons die of this dreaded disease every day...over 500,000 a year, with over 1,600,000 newly diagnosed with cancer each year. It is an epidemic! The accepted, conventional treatments of surgery, radiation and chemo- therapy offer no real assurance of cure, and the financial burdens these place on the patients, their families and the insurance companies are staggering – hundreds of billions of dollars per year. These medical costs are gradually bankrupting the country.

Thousands of dedicated researchers and clinicians are devoting their lives to the conquest and eradication of human cancer. Unfortunately, much of this money and these efforts are focussed on seeking new drugs and agents that will kill cancer; not on trying to discover or understand the fundamental causes or mechanisms underlying cancer. A primary reason for this is the premise that a chemical or biological "magic bullet" can be found that will magically destroy and cure cancer just as Fleming's discovery of penicillin prior to the second World War successfully healed most bacterial infections. If a chemotherapeutic agent temporarily kills some cancer cells, but does not cure the cancer, the patient will eventually die. Whereas, if medical research could discover the true underlying cause(s) of human cancer, the second most deadly health scourge in our lives next to heart disease could be cured or eliminated.

<u>*The Immune System and Cancer*</u>

Our body's first and best line of defense against disease is its own immune system. Cancer cells are characterised as the uncontrolled growth of our own mutant cells which escape immune surveillance and invade healthy tissues. Many researchers now believe that cancer growth closely parallels fetal growth within the mother's womb. In addition to extremely rapid cell division – common to cancer and fetal development – cancer cells share the same mechanism by which an embryo resists attack by its mother's immune system. Fetal cells produce the human growth hormone, human chorionic gonadotropin (HCG).

HCG, the first hormone produced in large quantities by the developing fetus, coats the surface of the fetal cells and shuts down attack by the mother's immune system. Cancer cells have been shown to also produce HCG, shielding them from recognition and attack and allowing them to grow and replicate in the body without stimulating an immunological response. A number of scientists now believe that a therapy that attacks and destroys HCG represents a potential key for successfully treating cancer.

Cancer has a number of contributing causes including bacterial, environmental exposure to toxins and carcinogens, stress weakening the natural immune system, nutritional deficiencies and excesses, as well as genetic and hereditary predisposition.

Largely overlooked and frequently ignored are the fundamental roles of bacteria and mycobacteria as a root cause of cancer.

Mycobacteria as an Underlying Root Cause of all Human Cancers

The late Dr. Virginia Livingston, M.D. (1906-1990) discovered and isolated a mycobacterium that is a primary cause of all human and animal cancers (a mycobacterium is a genus of rod-shaped acid-fast bacteria some of which cause Leprosy and Tuberculosis). To confirm that this isolated mycobacterium was indeed the root cause of cancer, she followed Koch's Postulate by injecting the mycobacteria into healthy animals that subsequently developed the same cancer as the donor in every case. She further demonstrated through brilliant research and clinical findings how this mycobacterium produces the many different forms of cancer and how it shields and protects itself from recognition and destruction by the natural human immune system.

Koch's Postulate: Nobel Prize recipient Robert Koch isolated and identified the Anthrax bacteria in 1876 and the Tubercular mycobacterium in 1883. In each case to prove that these bacteria caused the disease, Dr. Koch established the following criteria to satisfy his Postulate: That a single organism could be isolated and identified in every case of the disease; that the organism could be grown *in vitro* in pure culture; that the cultured organism would cause the disease when injected into healthy disease-free animals; and lastly, the organism could again be isolated and recovered from these now infected experimental animals.

Over the past 50 years, since her discovery of the cancer-causing mycobacterium, prominent research has established links between certain bacterial and viral infections, inflammation and most common cancers. Many investigators have reported the discovery of certain bacterial isolates from human and animal cancer tissue, as well as from the blood and urine of cancer patients.

Extensive evidence clearly linking this mycobacterium to cancer, was documented by Dr. Virginia Livingston (who defined it as the Progenitor cryptocides (PC) mycobacterium), Alexander Jackson, et.al., and published in the prestigious American Journal of Medicine Sciences in December 1950. Later confirmatory studies were published in the Annals and Transactions of the New York Academy of Sciences in 1970,

1972 and 1974.[1,2,3] Despite this convincing research, bacterial organisms have been generally ignored as a potential cause of cancer. The reason for this may lie in the evasive nature of the pleomorphic PC mycobacterium. It is a master of disguise, taking on many different forms in the course of its growth and replication cycles. (See Figures 1 & 2)

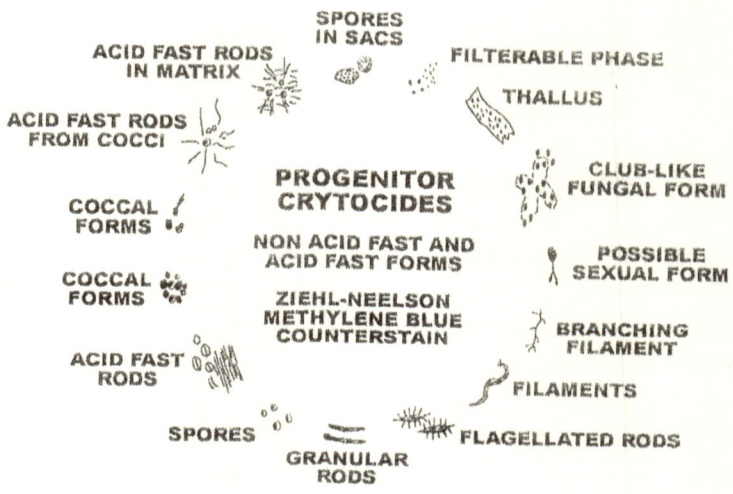

Figure 1: The Pleomorphic Forms of Progenitor Cryptocides

Mycobacteria

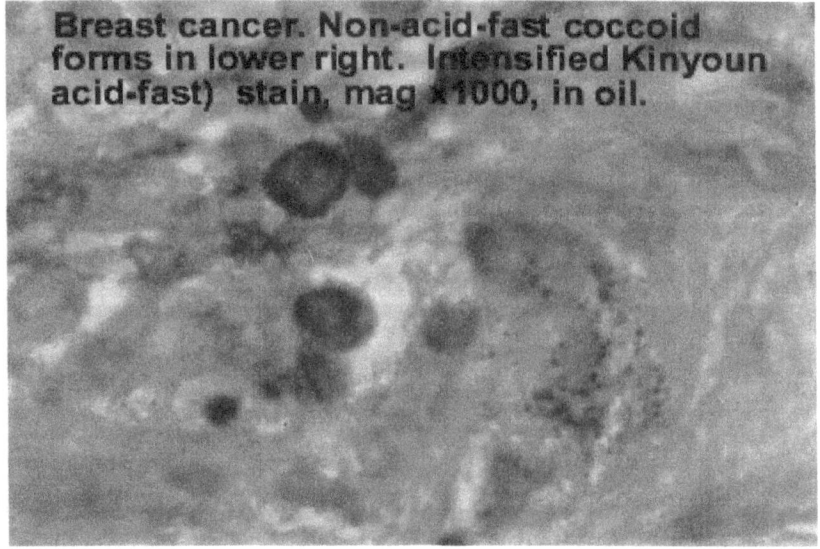

Figure 2: Cancer Mycobacteria in Breast Cancer (Courtesy of Dr. Alan Cantwell)

Dr. Livingston profoundly stated that all reproductive life, whether of the fetus or the cancer cell, is controlled by Human Chorionic Gonadotropin (HCG). **"It is the hormone of life"** (shielding the fetus containing the foreign male's sperm cells from destruction by the mother's immune system), and **"the hormone of death"** (shielding the cancer cells from detection and destruction by the cancer patient's immune system.) Her findings have been confirmed by numerous investigators worldwide, including Drs. Acevedo and Slifkin in 1976 at the Department of Laboratory Medicine of Allegheny General Hospital in Pittsburgh.*[4]

By demonstrating that all cancer cells possess PC mycobacteria and that these mycobacteria produce HCG, Dr. Livingston showed that the HCG growth hormone shields and protects these cancer cells from recognition and attack by the body's immune system. The fact that most normal cells do not produce HCG suggests that an anti-HCG immunotherapy may offer a very specific means of treating cancer. This therapy would enable the immune system to overcome the barricade shielding the cancer cell.

A number of scientists have come to believe that HCG represents an important clue for successfully treating cancer.

Because the PC mycobacterium in one of its many forms looks like a plant fungus, Dr. Livingston reasoned that a botanical substance might block its growth and replication. She discovered that CIS-14 (abscisic acid, a plant hormone called dormin) is a retinoid or Vitamin A analog that produces dormancy in seeds and plays a major regulatory role in the production of HCG by the PC microbe and the cancer cell. Cis-14 both blocks and deshields the HCG produced by the PC mycobacteria and the cancer cell. In turn, this allows immune system antibodies formed in response to the Immune Antigen (IA) vaccine to control and destroy both the PC mycobacteria and cancer cells.

She further discovered that the PC mycobacterium was closely related to the Tuberculin (TB) mycobacterium, and that a vaccine against one was cross-reactive to the other. Thus, Bacillus Calmette-Guerin (BCG) – a weakened live tubercular vaccine used to stimulate the human immune system's production of specific antibodies against Tuberculosis – was also an effective vaccine against human cancer.

Livingston Patient Outcome Survey Report

A survey study was conducted during October 2002 to determine how well the current cancer patient population at the Livingston Foundation Medical Center (LFMC) fared in terms of survival and quality of life. The patients were treated <u>only</u> with non-toxic immunotherapy to restore and enhance their immune systems to effectively fight against their serious chronic disease.

Of 532 patients treated and re-evaluated at LFMC in San Diego during the period 1999 through October 2002, 260 had been on Livingston immunotherapy for less than two years; 118 had been on the LFMC program for two to five years; and 126 patients had been on the program for more than five years (some for over 25 years) and are doing well. Their cancers have remained in remission and their quality of life has continued to be good.

Of the 324 patients with Breast, Prostate, Melanoma, Ovarian and Lymphoma Cancer (see Table 1 below), the Livingston immunotherapy program appeared to be particularly successful in restoring the immune sustem of these patients. In addition, the remaining 208 patients, representing 40 other types of cancer, also did well on LFMC immunotherapy; however, the number of patients in each cancer classification was statistically insignificant to allow them to be included in this limited survey study.

Despite the fact that many of these patients had been treated with surgery, radiation and/or chemotherapy before coming to LFMC with late-stage cancers, this survey indicated that Livingston non-toxic immunotherapy survival times and quality of life exceeded the so-called 'five year survival" standard adopted by conventional medicine.

TABLE 1 - LFMC PARTIAL PATIENT LONGEVITY SURVEY (10/02)

YEARS IN LFMC IMMUNOTHERAPY PROGRAM

Cancer Type Total		0-2	2-5	5-10	10-15	15-20	20-25	25-30	EXPIRED
BREAST	166	82	38	20	13	7	1	2	3
PROSTATE	73	26	22	15	8	2			
MELANOMA	17	5	2	4	2	2	2		5
OVARIAN	36	15	9	6	1				4
LYMPHOMAS	32	9	11	1	2	4	1		
TOTAL PATIENTS 324		137	82	46	25	16	4	2	12

TABLE 2 - LFMC PATIENT POPULATION SUMMARY OCTOBER, 2002 YEARS IN IMMUNOTHERAPY PROGRAM

CANCER TYPE	TOTAL #	0-2	2-5	5-10	10-15	15-20	20-25	25-30	EXPIRED
ABDOMINAL	1	1							
ADRENYL GLAND	1		1						
ANAL	3	2	1						
ANKLE	1					1			
BILIARY TRACT	1	1							
BLADDER	8	4	2	2					
BONE	3	2		1					
BRAIN	9	7							2
BREAST	166	82	38	20	13	7	1	2	3
COLON	53	38	5	3					4
CONNECTIVE TISSUE	7	3	3		1				
DUODENUM	1	1							
ENDOMETRIUM	3	1	1		1				
ESOPHOGEAL	3	2					1		
EYE	1			1					
GASTROESOPHAGEAL	1	1							
HODGKINS	2				1	1			
KIDNEY	8	4	2		1				1
LEUKEMIA	10	2	3	5					
LIVER	3	2			1				
LUNG	37	31	3	1					2
LYMPHOMA, MALIGNANT	16	6	5		3	1			1
MASTOID	1		1						
MELANOMA	17	5	2	4	2	2	2		
MULTIPLE MYELOMA	3		2	1					
NASOPHARYNX	1	1							
NECK	1		1						
NON-HODGKIN'S LYMPHOMA	14	3	6	1	1				3
NOSE	1	1							
OVARIAN	36	15	9	6		1			5
PANCREATIC	6	3							3
PHARYNX	2	1			1				
PROSTATE	73	26	22	15	8	2			
RECTAL	7	4	2						1
SALIVARY GLAND	1								1
SCALP	1		1						
SKIN	2		1	1					
STOMACH	7	4	2		1				
TESTICULAR	1		1						
THROAT	1				1				
THYMUS GLAND	2		1			1			
THYROID	4	3		1					
TONGUE	6	2	2	1		1			
UTERINE	8	2	1	5					
TOTALS	532	260	118	68	35	17	4	2	28
% OF TOTAL #	100.	48.9	22.2	12.8	6,6	3.2	0.7	0.4	5.3

2. The Role of Specific Vaccines in the Treatment of Cancer

Bacillus Calmette-Guerin

In 1973, Dr. Livingston learned about the amazing success Dr. Chisato Maruyama, M.D. (head of the Research Institute of Vaccine Therapy for Tumors and Infectious Diseases at the Nippon Medical School, Tokyo, Japan) had in treating both Leprosy and Cancer with Bacillus Calmette-Guerin (BCG). Dr. Maruyama used the BCG vaccine to effectively treat Leprosy (a relative of tuberculosis), and 871 cases of cancer of the lung, esophagas, stomach, intestine, liver, pancreas, uterus, ovary, etc. It proved very effective in 144 cases, exhibiting some effect in 271 cases, and little or no effect in the remaining 316 cases.

Dr . Livingston also learned that Dr. Maruyama had increased the strength of his BCG vaccine by almost 1,000 fold by stripping the protein sheath surrounding the active BCG DNA, which then proved far more viable. Armed with this knowledge, she asked her biochemical associate, Dr. John J. Majnarich, PhD, to strip the protein sheath from a patient's autogenous (self-) PC mycobacterium to produce a universal anti-cancer vaccine that could be widely used in treating all patients' cancers – thus, the Purified Antigen (PA) vaccine was created.

The PC Immune Antigen (IA) Autogenous Vaccine

The Immune Antigen (IA) is a specific antigen made from each individual cancer patient's strain of the *Progenitor cryptocides* (PC) mycobacterium. It is used to stimulate and enhance the immune system's ability to produce specific antibodies against the disease-causing PC. The IA is the keystone in stimulating the cancer patient's immune system's ability to control and destroy the invasive PC, the primary root cause of most human cancers.

The PC Purified Antigen (PA) Vaccine

The Purified Antigen residue fractions are prepared from a PC culture producing chorionic gonadotropin. The PA consists of the cell wall components and associated proteins and glycoproteins of a chorionic gonadotropin (HCG) producing organism isolated from human tumor tissue. By stripping the protein sheath from the mycobacterium DNA, the Purified Antigen became a universal vaccine, rather than an autogenous vaccine exclusively for the original donor. It could now be used as a universal vaccine against any patients' cancer. Antigens prepared from the PC mycobacterium activate and stimulate the patient's specific immune response to the cancer-causing mycobacteria. This effect was demonstrated extensively against a spectrum of laboratory animal tumors (see Figure 2).

Figure 2. Animal Tumor Response to Purified Antigen (PA) Vaccine

Tumor Model	No. of Mice	% Tumor Inhibition
Sarcoma 180	60	68 – 72
Sarcoma 91 (Melanoma)	160	56 – 87
T-241	50	34 – 58
Leukemia 1498	140	65 – 71
Lewis Lung Carcinoma	60	78 – 91
Leukemia P388	100	79 – 86
Rauscher Virus	40	95 – 100

Subsequent clinical trials, though limited in scope, provided strong evidence that the PA extract produced a significant response to many different neoplastic diseases and severe immuno deficiencies such as AIDS (see Figure 3).

Figure 3. Clinical Trial of Purified Antigen (PA) Vaccine on a Spectrum of Tumors

# of cases	Location & Cancer Type	Dead	Recurrent	Partial Remission	Remission
21	Breast w/metastases	2	2		17
5	Lung w/metastases	1			4
4	Uterine w/metastases			1	3
3	Ovarian w/metastase			1	2
6	Colon w/metastases	1	1		4
6	Melanomas	1			5
2	Basal Cell Carcinomas			1	1
3	Prostate			1	2
2	Kidney w/metastases			1	1
1	Pancreatic w/liver metastases				1
1	Pelvic				1
1	Esophageal				1
1	Larynx				1
6	Hodgkin's Lymphoma		2		4
TOTALS		**5**	**5**	**5**	**47**

The e± Factor Vaccine

The e+Factor is a non-toxic synthetic seven amino acid peptide developed by Dr. John J. Majnarich for the Livingston Foundation Medical Center. Alone, the e+Factor elicits a strong antigen-specific immune T-lymphocyte activity against systemic PC mycobacteria and cancer cells. In combination with the Immune Antigen or the Purified Antigen, it stimulates and exhibits strong anti-cancer activity (anti-angiogenesis).

In preclinical trials against Sarcoma 180 tumors in white mice, the e+Factor produced 77.3% inhibition of the tumors. When added to the Immune Antigen (IA), the combination vaccine of IA and the e+Factor produced 76.6% inhibition of the tumors, whereas the Immune Antigen (IA) vaccine alone only produced 44.6% inhibition of the Sarcoma 180.

The Custom Formula Vaccine

The Custom Formula is a splenic, thymus and liver extract from disease-free sheep used to stimulate white blood cell (WBC) production in the bone marrow of cancer patients. Often, after extensive cancer treatment, the patient's white count is extremely low, thus compromising the immune system's ability to mobilize immune cells and attack and destroy invading disease-causing bacteria and mycobacteria.

The Role of Certain Carcinogens and Viruses in Causing Cancers

Pesticides and Breast Cancer

The New York Times carried a front page article that states that two of the leading agricultural insecticides extensively used in the United States had been clearly linked to breast cancer. These were produced[*1] by Imperial Chemicals (ICI) of Great Britain and its subsidiary, Astra Zeneca. "Ironically, the United States Breast Cancer Month was founded and sponsored by Zeneca Chemicals (a subsidiary of ICI) which earns millions from the sales of carcinogenic pesticides, and now, as Astra Zeneca, from the breast cancer drug, Tamoxifen.. Zeneca also purchased the largest for-profit cancer treatment centers in the United States, Salick Health Care Inc., neatly assuring profits both from causing and curing breast cancer. Among a few of the many pesticides that increase the risk of breast cancer are: lindane, permethrin, cypermethrin, and captan.

"Most of the pesticides implicated in breast cancer are still in common use because of regulatory failure to update scientific research; instead, regulators rely on toxicological data suppied by the pesticide manufacturers as 'proof' that a pesticide is 'acceptable'. " [*5]

" Breast cancer incidence rates in the United States rose by 24% between 1973 and 1991. Breast cancer risk factors may vary by birth cohort, including age at menarche, age at first birth, physical activity, obesity, diet, alcohol intake, **estrogen therapy and exposure to environmental organochlorines (pesticides).** [*6]

H. pylori (Helicobacter pylori)[*7] – a Bacterial Cause of Cancer

H. pylori (Helicobacter pylori) bacteria infects over half the world's population. Ordinarily it causes no symptoms; however, it can lead to serious complications such as gastric ulcers and cancer.   H. Pylori is primarily passed from person to person through direct contact with saliva or fecal matter. Well adapted to the stomach's acid environment, it produces an enzyme creating a low-acid buffer zone. If untreated, H. Pylori can cause gastric cancer.

Human papillomavirus (HPV) [*8,9]

A type of Human papillomavirus is the most sexually transmitted infection in the United States. It is the primary cause of cervical cancer in women; and in men it can also lead to cancers of the anus and penis. A cervical Pap test is routinely used to detect abnormal cells. Subsequent surgical removal of precancerous lesions prevents the further development of HPV-induced cervical cancer. Since the Pap smear was developed, there has been a 70% reduction in the incidence of cervical cancer. Gardisil, a vaccine designed to prevent HPV infections in young women, has been shown to have side effects and is not effective in treating already infected patients.

Mesothelioma and Asbestos [*10]

Mesothelioma is a cancer almost always caused by exposure to asbestos dust and fibers. Cancer cells primarily develop in the pleura lining of the lungs and internal chest wall. Patients with mesothelioma have an extremely poor prognosis. Treatment with surgery, radiation and chemotherapy offer little help.

Cytomegalovirus and p53 Protein

Cytomegalovirus (CMV) is a member of the Herpes virus family which also includes the Epstein-Barr virus. Infecting over 50% of the U.S. adult population, it frequently remains dormant for many years and does not often cause disease. However, CMV frequently blocks the p53 tumor-suppressing protein, allowing the unchecked growth of cancerous tissue.   CMV can be transmitted from person to person through bodily fluids, and can also be transmitted sexually and via breast milk. A selective anti-Herpes vaccine must be developed so that the Cytomegalovirus can be destroyed and the p53 anti-cancer protein allowed to naturally suppress cancer cell proliferation and resultant tumor growth.

The p53 protein is extremely important in regulating the human cell cycle. The p53 protein not only supresses cancerous tumor growth and cellular development, but also prevents genome mutation.

3. The Role of Cigarette Smoke in Causing Lung Cancer

In 1962 I left the Children's Cancer Research Foundation (now Dana Farber) as Associate Director to work with Dr. Edwin Land on the development of instant color film. Because of my biochemistry training, I was also responsible for reviewing the toxicity testing of all chemicals used in the color film process at Arthur D. Little, a leading consulting firm in Boston.

On one occasion Dr. Charles Kensler, Senior Vice President in charge of the Life Sciences Division at Arthur D. Little, asked me to come into his laboratory to look at some important research results on the testing of a new charcoal filter for Liggett & Myers, a large tobacco company. I had first met Dr. Kensler at Sloan Kettering Institute for Cancer Research where at the time he was one of our senior Scientific Advisors and I was Assistant to the Scientific Director at SKI.

He showed me lung tissue slides of experimental mice that had been exposed sequentially to unfiltered cigarette smoke in a test chamber. Initially, the cilia in the mouse lung tissue showed signs of poisoning; then, as the remaining mice continued to be exposed to more smoke, the cilia became pre-cancerous lesions; and finally cancerous lesions. I was then shown a second set of tissue slides of mice sequentially exposed to cigarette smoke first passed through a charcoal filter – no poisoned cilia; no pre-cancerous lesions; and no cancerous lesions.[*11,12]

Dr. Kensler explained that at the burning temperature of the cigarettes (some 2000°F), some of the air (composed of roughly 4/5 Nitrogen and 1/5 Oxygen) was converted into Nitrogen Oxides of combustion. It was these oxides of combustion in the gaseous phase of the smoke that were responsible for the poisoning of the cilia and the formation of lung cancer. The charcoal filter removed up to 95% of these cancer-causing oxides.

One has to ask why the FDA doesn't require the tobacco industry to use charcoal filters on all cigarettes? The tobacco industry could easily and inexpensively incorporate charcoal filters on cigarettes thus eliminating a major cause of lung cancer.

4. The Role of Nutrition in Causing and Preventing Cancer

__Nutrition, Diet and the Immune System__

One often hears: "You are what you eat..." In the United States today, the pace of everyday life places us under a great deal of stress. Our lifestyles frequently find us eating on the run, catching a quick bite at a drive-through fast-food restaurant, or popping a frozen dinner in the microwave because it's easy and we're too exhausted or pressed for time to fix a balanced nutritional dinner.

Our food often comes from farms and countries hundreds or even thousands of miles away. It frequently reaches our tables only after detouring through processing plants, through complex machines devised to peel, cook and preserve fresh foods, transforming them into bagged, canned, frozen or dried products. Fruits and vegetables are sprayed with insecticides, waxed, pickled, sweetened, dyed, conditioned and sterilized, reducing the nutritional value to only a fraction nature intended by the time we eat them.

Unfortunately, our modern dietary habits tend to be deficient in the essential nutrients necessary to the maintenance of a healthy, vital immune system. Human cells, particularly those involved with the body's immune system, have highly specialized functions and require many nutrients to carry out their duties. As the immune system cannot be sustained by a diet of devitalized foods, poor nutrition sets the stage for serious illness. As a food group, unprocessed living plants, fruits, cereals and vegetables contain all the essential nutrients necessary for good health.

Raw or lightly cooked vegetables provide the enzymes and vitamins essential to a healthy immune system. It has been shown that foods rich in Vitamin A guard against chemical carcinogenesis. Vitamin C promotes healing and the building of strong coll- agen, the "glue" that binds us together. Nicotinamide, Vitamin B-12 and riboflavin increase cellular oxidation, and all assist in promoting a stronger, healthier immune system.

While we pride ourselves on being the "best fed" nation in history, our diets rarely provide us with all the nutrients needed by our bodies and our immune systems. It is perhaps the ultimate irony that we are responsible for crippling our own immune systems through nutritional starvation. The "best fed" nation in the world also has the highest incidence of cancer and heart disease, What we must realize is that what we eat effects our chances of getting cancer, and that there is a scientifically proven diet that guarantees fortification of the immune system against cancer and other serious chronic diseases.

Nutrition and Cancer

Dr. Virginia Livingston isolated and identified pleomorphic (existing in many forms) mycobacterium from human cancers that mutates into pathogenic (disease promoting) forms. These mycobacteria depend on a weak immune system which allows them to grow and produce a growth promoting hormone (HCG) that permits them to multiply and which shields them from identification and attack by the natural immune system. As the cancer grows stronger, so does the chaos in the body bringing about many changes involving nutrition.

We must first ask: Why is the immune system weak and how can it be strengthened? The majority of cellular components that make up the immune system are comprised of proteins. The immune system requires a steady intake of protein in order to manufacture its armies of specialized disease-fighting cells and antibodies. Proteins are made up of amino acids. There are 22 different amino acids which are arranged by the body in varying sequences to produce the required immune system proteins. Our bodies normally synthesize many of these; however, there are eight amino acids (the essential amino acids) that cannot be synthesized and must be derived from our diet. These are found scattered among certain plants and vegetables, or derived in a more concentrated form from various animal sources.

We discover that animal protein can provide a rich source of available components from which our bodies can readily draw needed protein. However, animal protein may also contain toxins and other pollutants; it is a growth medium for microbes and bacteria, many of which are pathogenic like the cancer-causing mycobacteria. This raises a serious risk-benefit question: Is the benefit of having a rich source of animal protein available for the immune system worth the risk of burdening the body with toxins, pathogens and pollutants?

Animal protein is only as good as the health of the animal itself. Government required food inspection is inadequate since it is limited to gross inspection and does not involve microscopic examination of tissues except in rare cases. As much as 25% of the cattle going to slaughter harbor infection by the cancer-causing mycobacteria. All chickens are a potential source of protein because their body temperature of 105°F eliminates many of the common pathogens such as staphalococcus, but unfortunately, provide an ideal incubation environment for the cancer mycobacterium. Chickens can be successfully innoculated with the Purified Antigen (developed by Drs. John Majnarich and Virginia Livingston) creating pathogen-free chickens and eggs. Since only one in five million sheep have cancer, lamb, wild game and fish from unpolluted waters provide excellent sources of protein.

An alternative source of protein may be derived from plants, particularly from beans, whole grains, raw nuts, seeds, and quality soy products as well as designer-type protein powders, many of which include concentrated greens. At first glance, one might compare total protein content and find plant sources higher. However, the proportion of the essential amino acids may be more limited. Therefore, a carefully selected variety of plant sources may provide the best bioavailability of protein for the human immune system, and particularly for one weakened by a debilitating chronic disease. Looking at the other constituents, plants provide and include a number of nutrients, and in their refined state, a generous amount of fiber. Fiber not only has an affinity for trapping toxins, pollutants, excess hormones and cholesterol, but also represents a food source for beneficial organisms that cohabit the body. This may be one of the strongest links in maintaining a symbiotic system (the intimate living together of two dissimilar organisms in a mutually beneficial relationship).

An enormous amount of the foods we eat, including our meat and vegetables, have been frozen somewhere along the line – either in a refrigerated truck or railroad car, or in the store itself. Food can lose many of its natural nutrients in the freezing and defrosting process. Washing foods to remove surface contaminants and preservative chemicals removes many minerals and vitamins found in abundance on the skins and outer layers of fruits and vegetables. Consider further: insecticides on fruit; preservatives to keep bread fresh; additives to make tomatoes and beets redder; additives used in processing cheese; sweeteners in catsup and soft drinks; sugar in cereals and peanut butter; and the list goes on and on – the chemical reactions of these agents with the remaining nutrients in our food can render the vitamins and minerals of little or no value. It's not easy to eat healthy, but with a little extra effort, it can be done (see Tables 1 & 2).

Table 1 – <u>Foods to Include in a Healthy Diet</u>

Avocado   Mash ¼ avocado with a fork and stir into a glass of freshly pressed juice.
Allow to remain for 15 to 20 minutes before drinking.

Beverages   Herbal teas, sesame or nut milk, cereal beverages (eg-Postum), chicory or dandelion teas.

Breads   Millet, rye, buckwheat, whole wheat, bran, corn, seven-grain, corn tortillas. Only whole grains freshly ground and free of preservatives.

Cereals   Millet, oatmeal, brown and wild rice, buckwheat, alfalfa, groats, barley, cornmeal, oatmeal, cracked-wheat and seven grain. Freshly ground, rolled flakes or whole grain only.

Desserts   Fresh whole fruits, fresh fruit cocktails, and natural fruit gelatin.

Fat/Oil   Extra virgin cold-pressed olive oil, avocado, raw butter, flax oil & unrefined coconut butter.

Fish   Fresh and salt water fish, broiled, baked or poached. Avoid shellfish.

Fruits   Fresh, organic fruits. Apples, apricots, bananas, berries, cherries, currants, grapes, guava, mangoes, melon, nectarines, papaya, peaches, pineapple, pears, plums, persimmons, tangerines, and unsulfured dried fruits. When possible, eat the seeds or kernels with the fruit.

Juices   Only freshly pressed juices and frozen pineapple juice. Also include beet leaves, chicory, escarole, Swiss chard, watercress, beets and cucumbers.

Meat	Lamb and its internal organs such as heart and extra fresh liver.
Milk	Substitute soy or rice milk, seed and nut milks such as sesame, sunflower, almond and cashew.
Nuts	Fresh raw nuts, particularly walnuts, almonds, cashews, and pecans. Raw nut butters freshly made in the blender or juicer.
Potatoes	Baked or steamed with jackets. Sweet potatoes are an important dietary addition.
Salads	Use raw fruits and mature vegetables (listed below), shredded or chopped, separate or combined, such as shredded apple and carrot.
Soups	Homemade soups from listed ingredients; barley, brown rice or millet can be added.
Sprouts	Sprouts to the age of seven days are permitted since Vitamin A and the abscicins are very rich in these. After eight days, the gibberelins are produced, which promote rapid growth of seedlings and tumors, and they neutralize the natural abscisic acid.
Vegetables	Organically grown raw or freshly cooked: artichokes, asparagus, beets, broccoli, carrots, cauliflower, chives, corn, endive, green and wax beans, kale, legumes, potatoes, spinach, squash, Swiss chard, watercress and any fresh seasonal vegetable.

Table 2 – <u>Foods to Absolutely Avoid</u>

Beverages	Alcohol, cocoa, coffee, milk and all soft drinks
Bread	White and blended breads made from white flour
Cereals	Processed cereals, puffed or sugared. No white rice.
Cheese	Processed cheeses
Desserts	Canned or frozen fruits, All pastries, gelatin, custards, sauces, ice cream, and candy (due to high sugar content which feeds cancer).
Eggs	Forbidden due to pathogenic mycobacteria in all chickens and eggs.
Meats	Beef, veal, chicken or pork in any form. Fried, smoked, salted or processed meats such as sausage or cold cuts.
Milk	Only whipping cream and raw butter may be used in moderation.
Nuts	Salted nuts.
Potatoes	Commercial French fries or potato chips.
Soups	Canned or frozen soups, fat stock, bouillion or dehydrated consomme.
Sprouts	Mature sprouts.

Sweets   White refined sugar and sugar products such as candy (for sugar fuels cancer). All sugar substitutes.

Vegetables   Sprayed and canned. Sulfur and high sodium foods. Frozen vegetables preferred to canned only when fresh are not available.

Avoid contact with all chemicals, cleaning solutions containing toxic chemicals, solvents, paint removers, insect sprays, petroleum products, deodorants, hair dyes, disinfectants, pest strips, perchloroethylene (dry cleaning solvents), and phthalates in fragrances & air fresheners (causing endocrine disruptions).

5. Iridodial, a Cancer Gene Reversal Agent in the Treatment of Cancer

The Iridodials*[17] are a primary source of dialdehydes, and are extremely powerful genetic-repair factors. The first pulmonary tumor regression was observed by Dr. Didier of Gifhorn, Germany. Insects, and particularly ants, have the capacity to produce large amounts of gene-repairing substances efficiently. Hence, these insects rarely develop tumors and are able to host unbelievable amounts of viruses without ill effect; yet, they have no immune system. Dr. Hans Nieper, MD, observed the iridodials are extremely effective in terminal breast cancer so long as the tumor has not grown beyond a certain size. In 1973 at the Silbersee Hospital in Hannover, Germany, Dr. Nieper witnessed a so-called spontaneous healing of this patient with advanced breast cancer. The cause of this and the healing of widespread bone cancer metastases was due to genetic repair in the cancer cells rather than the expected immune system response as with patients responding to a viral or bacterial infection.

During an activated immune response, the anti-bodies, which consist of peptides, are the elements actually fighting the bacterial and viral foreign invaders. In contrast, various steroid and chinoid structures have a primary role in the inner cell genetic repair system. There are over one hundred fifty chemical structures that possess reparative abilities against degenerative cancer cells, damaged cells and certain large viruses such as the herpes-type virus.

The chromosomal genetic systems of the human cell contain approximately two billion base pairs, constituting a "computer" that has the extremely high capacity of two billion bits. Most of the stored information is securely sealed within a locking mechanism. Only a small amount of the stored genetic information may be released at any time. Because of the strict regulation of the genetic system, the preservation of the type and function of the genes is pre-programmed and secure.

Under certain conditions, previously secure gene systems will begin to release information pertaining to the whole cell, which eventually will lead to chaos. Depending on the frequency of unsolicited released genetic information, a cancer cell can be formed; and out of that cancer cell, a malignant tumor can be formed. The gene responsible for causing such chaos is called an "oncogene" as defined by Peter Duesberg of Berkeley, California. In addition to the oncogenic genes, the lipids (fatty substances that serve as cellular food and are important constituents of cell mambranes) of the mitochondrial membrane of the cell can change into malignant lipids and are also reponsible for the formation of cancer (Malignolipin). It is essential to keep the oncogenic genes latent and ineffective. Because of this, "anti-oncogenic" genes are gaining special interest.

There are two basic gene repair mechanisms focussed on extinguishing oncogenic genes and possibly restoring cancer cells to their normal function, or, disposing of them. They are: the anti-oncogenic genes in the gene system; and the so-called oncostatins (peptides) in the cell plasma. These oncostatins require the healthy condenser function of the outer membrane (50 to 90 kilovolt per centimeter). Healthy embryonic cells have the potential to re-program cancer cells back to normal cells.

Iridodials are more effective and significantly less expensive than other gene reversal modalities. They are also non-toxic and completely free of side effects. The future clinical use of iridodial would become considerably less expensive than chemotherapy, and special oncology clinics would be unnecessary. Cancer patients could be treated successfully and less expensively by their family physician or by an internist rather than an oncologist.

6. Laetrile, a Promising Non-toxic Anticcancer Agent, Banned in the U.S.

In 1920, Dr. Ernst Krebs, MD, prepared an alcohol oral extract from finely ground apricot kernels which he later freeze-dried into a yellow powder. He found it to be quite toxic but effective against a spectrum of cancers. His son, Ernst Krebs, Jr., PhD, later developed a less toxic extract which he called Laetrile(*18,19) for Laevomandelonitrile. In the process of removing the yellow powder to create the pure white crystalline Laetrile, he unknowingly removed abscisic acid, an important compound necessary to deshield the cancer cells from attack and destruction by Laetrrile.and the host's natural immune system antibodies.

The Laetrile molecule consists of a single glucose molecule bonded to a molecule of HCN (hydrocyanic acid), and benzaldehyde (an aspirin-like pain liller). The glucose mandelonitrile compound can <u>only</u> be split, releasing the toxic HCN, at the membrane surface of cancer cells by **rhodanase** an enzyme produced only by cancer cells, thus selectively destroying the cancer cells and not normal cells. The benzaldehyde, a powerful pain killer and nontoxic anticancer substance, is also released primarily at the cancer cell site.

During the early 1970's, the American Cancer Society, the National Cancer Institute, the American Medical Society, the powerful FDA and numerous state medical societies conducted a widespread campaign against Laetrile. They claimed Laetrile had no merit as an anti-cancer substance and that the HCN would poison cancer patients, purposely ignoring the fact that the HCN when released at cancer cells would be completely metabolized and excreted harmlessly from the body. Many Laetrile proponents believed the Laetrile ban was in the financial interest of the large pharmaceutical companies producing the expensive, toxic chemotherapy agents; particularly since Laetrile, as a natural substance, offered little economic return, could not be patented, and could be crudely made in most American kitchens.

Since the ban made it difficult for Dr. Nieper to obtain Laetrile from the United States, he asked his close friend and collaborator Dr. Franz Kohler, Sr.(the discoveror of acrylic acid and inventor of Plexiglass) to synthesize a series of mandelonitriles substituting various chemical compounds for glucose in the mandelonitrile molecule. From this series Nieper submitted three compounds: the ureyl-, nicotinyl-, and para-amonobenzole-mendelonitriles, to Sloan-Kettering Institute in New York for anticancer testing. These showed great promise as potential nontoxic anticancer agents. Nieper subsequently used a mixture of the ureyl- and nicotinyl-mandelonitriles in his clinical practice with great results. Further, these synthetic mandelonitriles also proved to be effective gene-repairing substances.

In 1973, Alecia Buttons, the wife of the comedian Red Buttons, went to Dr. Nieper for treatment of her terminal cancer. Following extensive treatment with the new Mandelonitriles, her cancer went into stable remission. She was alive and cancer free in the 1990's and credited the treatment of her cancer with the mandelonitriles with saving her life.

7. The Role of Mycobacteria and Viruses in Cardiovascular Disease

Mycobacteria and Viruses in Cardiovascular Disease

Dr. Virginia Livingston barely survived an initial heart attack in 1962. Hospitalized for 10 days, and after declining an arterial bypass was confined to her home for the next two years. During this time she reviewed her past scientific research papers on cancer and mycobacteria. She noted that many of the research animals innoculated with the cancer mycobacteria also developed heart lesions. A number of clinical researchers in England reported finding strange microbial bodies in the hearts of patients who had died of coronary disease.

A close friend and colleague, Dr. Eleanor Alexander-Jackson, offered to prepare an autogenous vaccine made from cultured dormant cancer mycobacteria remaining in Dr. Livingston's system from an earlier successfully treated cancer of the forehead. This vaccine effectively controlled the mycobacteria responsible for her heart attack and restored her heart muscle.

In 1965, Drs. Livingston and Alexander-Jackson published a paper[*13] in which they proposed a theory that there are microbial bodies in the lesions of heart diseases and that they are particularly numerous in areas where blood vessels have ruptured. The prevailing theory in the 1970's suggested that coronary blood vessels are narrowed due to arteriosclerosis, and that the primary factors in this type of heart disease are cholesterol and fatty deposits in the walls of the vessels and being overweight. More recently, medical researchers are becoming aware that the blood vessels themselves are not so much involved as the suporting tissues and muscles of the heart so that rupture of these vessels is due to extrinsic factors outside the vessel rather than from intrinsic disease.

Dr. Joseph L. Melnick, a virologist at Baylor College of Medicine in Houston, suggested that the Herpes family of viruses may trigger coronary heart disease. He reasoned that many people are infected early in their lives with one or more of the Herpes viruses which may remain dormant for years. Melnick found evidence that cytomegalovirus (CMV), a member of the Herpes family of viruses, along with cholesterol and a fat-rich diet may initiate atherosclerosis by promoting plaque formation in arterial walls.

In 1964 Drs. Stephen Epstein and Edith Speir of New York proposed a link between arterial disease and cancer – the p53 gene – an anti-tumor gene "whose loss or inactivation may contribute to as many as 50% of all human cancers." In 1997, Dr. Epstein confirmed that the cytomegalovirus in heart patients was responsible for blocking the action of the vital growth-limiting p53 protein. When cytomegalovirus combines with the P53 protein and blocks its normal function, the tissues at the surgical site continue to grow, and after eight to ten years, seriously constrict blood flow through these tissue-clogged arteries. In a similar fashion, cytomegalovirus also inactivates the p53 protein at tumor sites allowing unchecked growth of the cancerous tissue.

Cytomegalovirus defined (From Wikipedia, the free encyclopedia)

Cytomegalovirus 5 (HCMV) is a herpes virus strain infecting 50% - 80% of the adult United States population. Infectious HCMV may be found in urine, blood, semen, tears and breast milk. Most healthy people who are infected by HCMV after birth have no symptoms; some develop an infectious mononucleosis/glandular fever-like syndrome. Initial HCMV infection, which is asymptomatic is followed by a prolonged, inapparent infection (dormancy) during which the virus resides in cells without causing detectable damage or clinical illness.

A recent study links CMV infection to high blood pressure in mice, and suggests that the CMV infection of blood vessel endothelial cells (EC) in humans is a major cause of atherosclerosis.

Hans A. Nieper, M.D. of Hannover, Germany, had flown to the United States in 1989 to lecture at several major cancer conferences. He arrived with a serious viral pneumonia and instead of returning to Germany immediately, he insisted on completing his scheduled speaking tour. On returning to Germany a week later, he relapsed and was hospitalized for some time and placed on strong antibiotic IV's. About three weeks after his recovery from the viral pneumonia, he suffered a serious heart attack. Coincidentally at that time, I had learned about the activity of the Cytomegalovirus and alerted his attending physicians. They immediately checked him for possible viral involvement, and discovered that the viral pneumonia had activated his long-dormant childhood varicella (chicken pox) virus which had violently attacked his heart. With this information, his physicians were able to treat him successfully. However, because he ignored his physicians' advice to limit his practice and to take it easy, he never fully recovered from this viral attack and sadly died of a fatal heart attack about three months later.

More clinical research must be undertaken to find and develop suitable vaccines for the Herpes family of viruses which are the causative agents in a number of other medical conditions such as the Herpes Zoster virus causing Shingles, which attacks sensory nerves causing severe pain; Bell's palsy, which causes localized edema and paralysis of the cranial and trigeminal facial nerves; and, as discussed earlier, the Cytomegalovirus, which by blocking the vital anti-cancer activity of the p53 protein leads to the proliferation and rapid uncontrolled growth of cancer cells and to fatal heart attacks.

8. The Role of Aluminum in Causing Multiple Sclerosis

The late Dr. Hans Nieper, M.D. of Hannover, Germany treated over 3,100 MS patients from all over the world; including over 2,400 from the United States. As a result, he was able to explain the origin and fundamentals of the disease.[*14]

"Multiple sclerosis is a progressive, degenerative disease of the central nervous system. It involves destruction of the myelin sheaths around nerve axons, resulting in scar tissue called *plaques* and ultimately destroying the nerves and their ability to function. MS patients experience membrane deterioration known as sclerosis.

"The myelin sheath is a complex membrane layer that is wound around a nerve fiber, referred to as the central axon (See Figure 5.1). This multilayered sheath is comprised of from five to thirty layers. Each individual layer of the laminated leaf is structurally identical with the membrane of a cell. This means that it has the ability to hold an electrical charge of opposite polarity, thereby functioning as an electrical condenser. It was only learned recently that this multi-layered condenser system acts as an electrical shunt to the central axon.

"The myelin sheath, which surrounds the nerve fiber like insulation around an electrical wire, is itself surrounded by the *medullary sheath* or *oligodendroglia,* which is composed of oligodendrocyte cells. These specific cells secrete *myelin* – a lipid or fatty substance that makes up the myelin sheath.

"During MS attacks, the myelin sheath is partially destroyed or demyelated by *killer T-cells* – powerful components of the immune sustem – leaving patches of scar tissue in the myelin sheath. This scar tissue interrupts communication being sent from the brain and nerve terminals, resulting in various disturbances of the nervous system, such as poor nerve signal transmission, bodily weakness and similar functional problems.

"Multiple sclerosis is a general membrane disease, not a nerve cell disease. It is an autoimmune disease that attacks all of the membranes, including cellular membrane degradation that affects the actual nerve fiber transmission. Many MS patients can't walk and are relegated to wheelchairs. They frequently suffer from degradation of the bones and organs, such as the adrenals.

"MS appears to be more common in temperate climates (1 case per 2,000) than in the tropics (1 case per 100,000) and in large dairy-producing regions of the world. The average onset age occurs between the ages of twenty and forty years, and women are more commonly affected than men. The primary factors causing the onset of MS are: aluminum and other elements, dairy products, heredity and viruses.

FIGURE 5.1

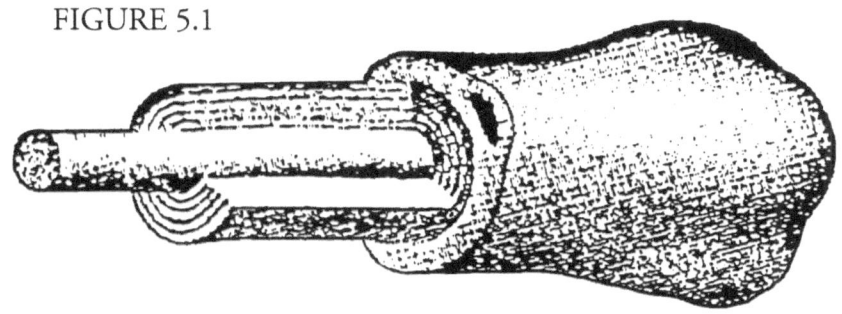

Aluminum as a Cause of Multiple Sclerosis

"Aluminum is strongly suspected as causing damage to membranes and the nervous system. A study of amyotropic lateral sclerosis (ALS), also known as Lou Gehrig's disease, in Guam revealed a very high incidence of ALS among aluminum welders. Dr. Nieper found evidence of aluminum exposure in his ALS patients as well. ALS is a disease in which brain-stem nerve cells are damaged by components of the herpes virus family, including the cytomegalo, Epstein-Barr, herpes II, and varicella viruses. There are two significant similarities between ALS and MS. The two diseases share similar motor symptoms, such as weakness, problems with walking, and tremors. Both are progressive diseases that can lead to the threatening *medula oblongata deficiency* symptoms. ALS originates with nerve dysfunction; the symptoms of MS result from faulty nerve dysfunction; the symptoms of MS result from faulty nerve transmissions that have taken place due to membrane sclerosis.

"Aluminum, in addition to influencing MS and ALS, may influence Alzheimer's disease as well. It has been established that the brains of Alzheimer's patients have ten to thirty times as much aluminum as the normal brain. We are exposed to the detrimental effects of aluminum constantly in pots and pans, in aluminum foil, through aluminum hydroxide in many deodorants and in U.S. soda-pop cans. It is best to avoid the use of products encased in or exposed to this element.

Dairy Products as a Possible Cause of MS

"MS is most prevalent in dairy producing regions worldwide. Two theories explain the close relationship between dairy consumption and MS. The first assumes that there are virus particles in milk that act as "starter viruses" and bring about the disease. The second originated and was researched in England about thirty years ago: It involves glutens, immune-active sugar-albumin complexes found in milk (and possibly cereals), which can also activate the condition so that it becomes clinically evident.

Heredity as a Possible Underlying Cause of MS

"The bodies inability to provide adequate amounts of aminoethylphosphate(AEP) to the cellular membranes may be due to some inherited predisposition. AEP is necessary to bind the proper electrical charge of the cell membrane. Incorrect membrane charge can lead to multiple sclerosis. This phenomena has been noted in mother and child and in identical twins. Supplementing the body with AEP should aid in correcting this deficiency.

Viruses as a Possible Cause of MS

"In order to understand the onset of MS, we need to know the underlying cause of immune aggression that works against the myelin. This autoimmune process is undoubtedly initiated by a viral infection.

Initially, it starts out as a healing action to destroy bacteria and other foreign protein invaders; but later, it somehow develops properties of its own, which, after a latency period, becomes programmed not only to destroy the initial virus, but also to attack the myelin membrane structure and perhaps the oligodendroglia that forms the myelin.

"There are numerous viruses suspected. The most significant virus seems to be measles. This was revealed by Dr. Mannweiler of Hamburg, Germany who stated 'Everyone, every patient with MS – and I mean 100 percent – had a severe bout with measles, or had been exposed to a measles-type virus.' A significant number of Americans became sick with MS in 1978. A very high percentage of these had swine flu immunization in 1977. Another equally important 'starter' virus is thought to be canine distemper. Additional suspected viruses are those of the mumps and chicken pox (varicella).

Orthodox Therapies for MS

"In the 1970's, a smear cure of a mercury salve was recommended in an attempt to block the autoimmune MS process. Some clinical effects were reported, but the prevalent side effect of kidney damage was extensive and the therapy was abandoned. Then, *cyclophosphamide* (Cytoxan, Endoxan) was occasionally prescribed in highly toxic doses, causing patients to lose their hair and experiencing severe bone-marrow damage to the immune system.

"A less toxic chemically-related *trophosphamide* (Ixoten) as an immuno-suppressive substance that is far better tolerated over a longer period of time. However, even Ixoten is only useful for a limited period of time.

"A medication recommended today for its immune-inhibitive ability is *azathioprine* (Imuran, Imunex). It is important to warn patients that azathioprine will cause liver damage if used for any length of time. In addition, there is an increased susceptibility to viral infection and possibly even cancer. The toxic effects of this therapy must be carefully considered.

"Prescribing *ACTH* (adrenocorticotropic hormone) to MS patients has become a widespread bad habit. ACTH stimulates the adrenal cortex to secrete minerals and glucocorticoids which, in turn, control chemical constitution of body fluids, metabolism, and secondary sexual characteristics. While there is often temporary improvement, the long-run result of ACTH use is one of steady deterioration. If ACTH is given for a short period of time, it is absolutely necessary to supply foods required by the adrenal cortex systems – raw foods, vitamin D_2, vitamin C (in large doses), betacarotene, and especially selenium.

An Alternative Medical MS Treatment Program

"Dr. Nieper created a successful alternative approach to the clinical treatment of Multiple Sclerosis. It involves a commitment to healthy lifestyle habits – good diet, regular exercise and a clean environment – and to the regular use of alternative orthomolecular substances.

Diet, Exercise and Environment

"A good diet is of extreme importance. Avoid milk and milk products as far as possible. An exception is allowed for natural French cheese, in which the glutens have been broken down by fermentation. A strict diet of raw, organically grown foods is best. When cooking and eating, avoid using aluminum cookware, avoid beverages stored in aluminum cans, and avoid using aluminum foil when cooking, roasting or storing foods.

"In addition to this dietary advice, controlled exercise and plenty of rest are essential. Active and passive (second-hand) smoking are strictly forbidden. The *nicotine effect* is mainly brought about by the smoke impairment of the electrical properties of the cell membrane. Discovered by Henri Laborit, French scientist, this exposure to poisonous smoke should be carefully avoided.

Alternative Substances for Treatment of MS

"The most important part of Dr. Nieper's treatment of MS is an attempt to correct the chemical and electrical defects of the cell membrane. The remedies of choice are the 2-AEP salts (colamine phosphate salts, including Calcium-AEP, and the Calcium, Potassium, Magnesium 2-AEP Complex). The ability of the Calcium 2-AEP to bind electrical charge on the myelin sheath membrane is a special physiological quality. Also, Calcium aspartate produces a similar effect. Some patients have been successfully treated solely with calcium aspartate for over thirty years.

"There are certain chemical substances necessary for the correct bonding of the electrical charge to the cell membrane of the myelin sheath. One of the most necessary is *aminoethylphosphate* (AEP)[*16]. This substance was first described by the famous American biochemist, Dr. Erwin Chargoff, in1940[*15]. If there is insufficient AEP in the cell membranes, the binding of the electrical charge and the electrical condenser function will be seriously impaired.

"To patients with autoimmune disorders, the body does not produce enough AEP to sufficiently supply blood and urine tract cells. Thus, these cells cannot maintain cellular integrity and function; for example, the electrostatic charge of the urinary tract cells is insufficient; thus, the electrostatic filter responsible for keeping the urinary tract clean, does not function adequately resulting in urinary tract infection.

"The combination of AEP insufficiency, functional membrane inferiority, and the resulting harm done by immune lymph cells and antibodies leads more or less to total destruction of the myelin through demyelization disease. This affects and impairs the nerve cells of the spinal column, resulting in loss of muscle control.

" In 1968, the German Health Authority (equivalent to our FDA) declared calcium 2-AEP an official medication in the treatment of multiple sclerosis. In 1986, Dr. George Morrisette conducted an extensive retrospective study on the effect of calcium 2-AEP on over 250 American patients. His findings revealed a positive response rate in MS patients of about 80 percent, supporting results experienced by Dr. Nieper in Germany.

"The Nieper MS therapy results in fairly reliable improvement – at least partly back to normal – of the bladder function, the intestinal sphincter muscles, and voluntary control of the big toe, even with badly crippled patients. In addition, the upper body functions are also improved, lessening the following: vertigo; slurred speech; loss of facial expression control; loss of motor function of the arm and hand; and especially the *medula oblongata deficiency symptoms*. This last group of very dangerous symptoms includes the inability to swallow, defective breathing functions, and poor regulation of circulation. These symptoms also occur with ALS and may be handled with continuous protection offered by the mineral salts of colamine phosphate (2-AEP) as well. Unfortunately, the disturbed motor functions of the upper thigh muscles, those essential in walking, are quite resistent to the Nieper therapy; improvement has been noted in only a few cases.

"Only eight bone fractures have occurred among 3,500 patients treated with calcium 2-AEP over a 30 year period. Further, a group of MS patients first seen and treated by Dr. Nieper in 1968 have not experienced any further progression of their disease and continue to follow his strict treatment and advice. "

BIBLIOGRAPHY

1. Acevedo, H.F., et. al.., "Human Chorionic Gonadotropin in Cancer Cell Systems," Press: Proc. Third Symposium for the Detection and Prevention of Cancer. Ed. H.E. Neiburgs, Marcel Dekker, Inc., New York, N.Y. (1976).

2. Livingston, A.M. M.D., Livingston, V. W.-C., M.D., Alexander-Jackson, PhD, and Wolter, G.H., PhD, "Toxic Fractions Obtained from Tumor Isolates and Related Clinical Implications." Annals of the New York Academy of Sciences, Vol. 174, Article 2, pp675-689 (October 30, 1970.)

3. Virginia Wuerthele-Caspe Livingston and Afton Munk Livingston, "Demonstration of Progenitor Cryptocides in the Blood of Patients with Collagen and Neoplastic Diseases." University of San Diego, San Diego, California 92110. (January 28,1972).

4. Virginia Wuerthele-Caspe Livingston, "Some Cultural, Immunological, and Biochemical Properties of Progenitor Cryptocides." Transactions of the New York Academy of Sciences, Series II, Vol. 36, No. 6, pp569-582 (June 1974.)

5. Muriel Watts, "Pesticides and Breast Cancer". PAN Magazine (Spring 2008).

6. King, S.E. and Schottenfeld, D. "The 'epidemic' of breast cancer in the U.S.--- determining the factors."Oncology (Hunting10(4) pp453-62; discussion, pp462, 464, 470-472. Department of Epidemiology, University of Michigan School of Public Health, Ann Arbor, USA.(April 1996).

7. Mayo Clinic Staff "H. pylori infection "www.mayoclinic.com/health/h.pylori/ DS00958(6/23/2009)

8. Schiffman, M. and Castle, P.E. "Human papillomavirus: epidemiology and public health". <u>Archives of Pathology&Laboratory Medicine 127(8)pp930-4</u>(2009.

9. Walboomers, J.M., Jacobs, M.V., Manos, M.M., *et al.* "Human papillomavirus is a necessary cause of invasive cervical cancer worldwide". <u>J. Pathol. 189</u> (1999).

10. Wikipedia, the free encyclopedia. <u>http://en.wikipedia.org/wiki/Mesothelioma</u>. pp1-13 (2009).

11. Statement of Dr. Charles J. Kensler to the Consumer Subcommittee of the Senate Committee on Commerce.http://tobaccodociments.org/bw/619009html.pp1-5(1967.

12. National Cancer Advisory Board, Meeting of the Ad Hoc Committee on Smoking and Health, Bethesda, Maryland. http://tobacco<u>documents.org/bw/133546.html</u> pp 1-5(June 17,1973)

13. Virginia W,C. Livingston, M.D. and Eleanor Alexander-Jackson, Ph.D."Mycobacterial Forms in Mycocardial Vascular Disease, <u>Journal of the American Women's Association, Vol.20, No.5,pp 449-52.</u> (May 1965)

14. Nieper, H.A., Alexander, A.D. and Eagle-Oden, G.S. <u>The Curious Man.</u> Chapter 8. pp115-130(1999)

15. Chargoff, Erwin and Keston, A.S. "2-Aminoethylphosphate, <u>Journal of Biological Chemistry, Vol 134,</u>p515(1940)

16. Nieper, Hans A. "A Clinical Study of Calcium Transport Substances Ca-*l,dl*-Aspartate and Ca-2-Aminoethylphosphate as Potent Agents Against Auto-immunity and other Anticytological Aggressions,"<u>Agressologie, Vol VIII, No.4,</u> pp4-16(1967)

17. Nieper, Hans A., M.D. " 'Iridodial' An Insect Derived Genetic Repair Factor of Important Antimalignant Effect", <u>Raum & Zeit</u>, (1990).

18. Culbert, Michael L. Vitamin B17 Forbidden Weapon Against Cancer, The Fight for Laetrile, <u>Arlington House Publishers</u>, New Rochelle, N.Y.(1974).

19. Nieper, Hans A., M.D., Arthur D. Alexander III, and G. S. Oden, The Curious Man, The Life and Works of Dr. Hans Nieper. <u>Avery Publishing Group</u>, Garden City, Long Island, New York, Chapter 3, pp 38-41(1999).

SUGGESTED READINGS

Alan Cantwell, Jr.M.D., *The Cancer Microbe*, Aries Rising Press, P.O. Box 29532, Los Angeles, CA 90029, 1990.

Alan Cantwell, M.D., *Four Women Against Cancer*, Aries Rising Press, P.O. 29532, Los Angeles, CA 90029, 2005.

David J. Hess, *Can Bacteria Cause Cancer?*, New York University Press, New York & London, 1997.

Ana Marie Canales, *The Cookbook*, The Livingston Foundation, St. Helena, CA , 1996.

Arthur D. Alexander III, *Livingston Immunotherapy*, 1st Books Library, Paperback ISBN: 1-4107-4012-9, 2003.

Micki Voisard, *Cancer Then Healing*, Stray Dog Press, P.O. Box 1099, Calistoga, CA 94515, 1998.

Arno Karlen, *Biography of a Germ*, Pantheon Books, a division of Random House, Inc., New York & Toronto, 2000.

Sunil K. Lal, Editor, *Biology of Emerging Viruses*, Annals of the New York Academy of Sciences, Vol 1102, (Published in Boston, MA), New York, N.Y., 2007.

Lewis Thomas, M.D., *The Lives of a Cell: Notes of a Biology Watcher*, The Viking Press, New York, N.Y., 1974.

Ralph W. Moss, *The Cancer Syndrome*, Grove Press, Inc., New York, N.Y., 1980.

ABOUT THE AUTHOR

Arthur Douglass Alexander *graduated in1952 from Case Institute of Technology where he majored in Chemical Engineering and Engineering Management. He later completed graduate studies in Biochemistry at Cornell Medical College (Sloan-Kettering Biosciences Division).*

Alexander has over 40 years experience in chemical and biological research and management activities. He served as Executive Asssitant to Dr. C. Chester Stock, Scientific Director of Sloan-Kettering Institute for Cancer Research in New York where he managed the Experimental Chemotherapy program at SKI's Walker Laboratory before being appointed Associate Director of the Children's Cancer Research Foundation in Boston.

He next worked with Dr. Edwin Land at Polaroid Corporation on the development of Polaroid's instant color film introduced in 1962. This was followed by a position as Senior Scientist in the Long Range Planning and Mission Analysis Division of NASA under the direction of Dr. Werner vonBraun. During this time Alexander met and became Scientific Advisor to the late Dr. Virginia Livingston in San Diego.

Currently on the Board of Directors of the Livingston Foundation for Cancer Research and Allied Diseases. He formerly served as Vice President, Assistant to the President, Chief Opererating Officer and Scientific Director of the Livingston Foundation Medical Center in San Diego.

Alexander is a member of many professional societies including the New York Academy of Sciences, the American Chemical Society, Royal Society of Chemists and is a fellow of the American Institute of Chemists. He has written many scientific papers, and Authored in 1998 The Curious Man, the Life and Works of Dr. Hans A. Nieper,MD, *of Hannover, Germany, and in 2002* Livingston Immunotherapy, The Treatment of Chronic, Immune-Deficient and Autoimmune Diseases.

www.ingramcontent.com/pod-product-compliance
Lightning Source LLC
Chambersburg PA
CBHW050336290526
45785CB00006B/2514